Science

Practice for the New York State Elementary Science Program Evaluation Test

Grade 4

Harcourt
SCHOOL PUBLISHERS

Visit *The Learning Site!*
www.harcourtschool.com

Printed in the United States of America

ISBN 0-15-353515-6

1 2 3 4 5 6 7 8 9 10 912 15 14 13 12 11 10 09 08 07 06

Contents

Learning About New York Elementary-Level Science

The next two pages give examples of the types of items found on the science portion of the New York State Elementary Science Program Evaluation Test.

About Multiple-Choice Items

Many of the items on the New York State Elementary Science Program Evaluation Test are multiple-choice. Each multiple-choice item has four answer choices. The tips that follow will help you answer these questions.

1. Read the question carefully. Restate the question in your own words.

2. Watch for key words such as *best, most, least,* or *except.*

3. The question might include tables, graphs, diagrams, or pictures. Study these carefully before choosing an answer.

4. Find the best answer for the question. Fill in the answer bubble for that answer. Do not make any stray marks around answer spaces.

1 Look at the diagram of the water cycle.

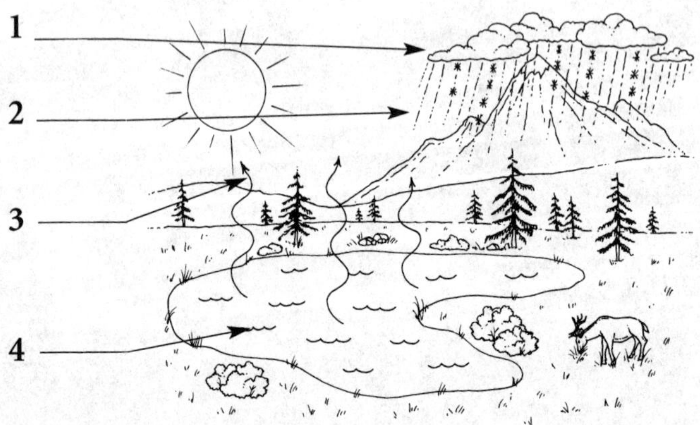

Which point shows the process in which liquid water is changed into a gas?

Ⓐ 1

Ⓑ 2

Ⓒ 3

Ⓓ 4

Learning About New York Elementary-Level Science

About Constructed-Response Items

For some items, you must write a brief answer to explain a science concept or to apply a science process skill. To receive the highest score answers should

- be complete.
- show understanding of the science content and processes.
- be accurate.
- communicate the ideas clearly.

---

1 Examine the food chain below involving grass, a prairie dog, and a hawk.

Grass → Prairie dog → Hawk

(a) Which consumer is an herbivore and which is a carnivore? Explain.

____

____

____

(b) How do all of the living things in the food chain get the energy to live from sunlight?

____

____

____

About Practice Sets

Each Practice Set consists of three multiple-choice items, 1 two-point short-answer item, and 1 four-point short-answer item. Some items have Tips, which give clues about how to answer the items. An item may have a graph, table, picture, or diagram. Study these carefully before answering the items.

Name ______________________________

Date ______________________________

Tip
Recall what a hypothesis is. Then eliminate any answers that do not test a hypothesis.

1 What do scientists do to test a hypothesis?

Ⓐ ask a question

Ⓑ communicate their results

Ⓒ make a prediction

Ⓓ plan and conduct an experiment

---

2 In the scientific method, which step occurs last?

Ⓐ make observations

Ⓑ communicate results

Ⓒ form a hypothesis

Ⓓ conduct an experiment

Name ______________________ Date ______________________

3 Which tool can you use to measure a volume of water?

Tip
Keep in mind that the tool must be able to measure the volume of the liquid.

 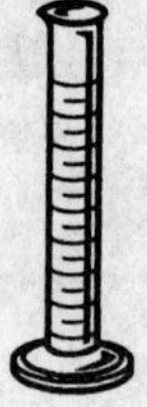

 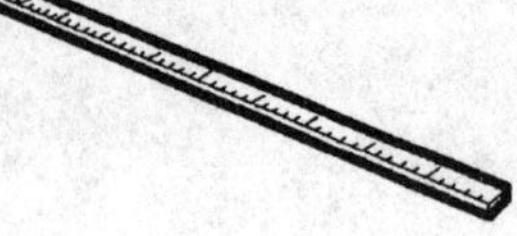

 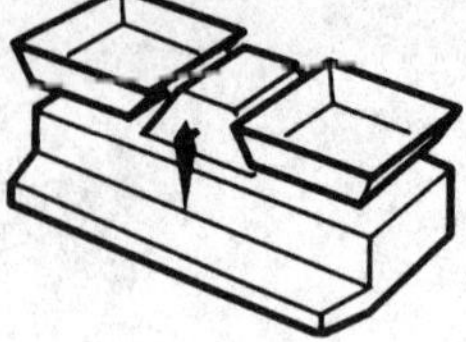

 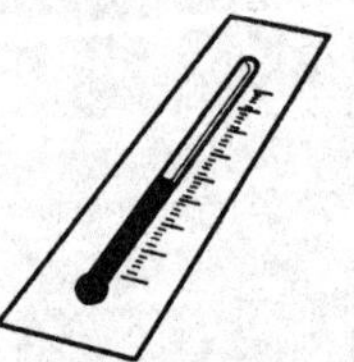

Name ______________________ Date ______________________

4 A student measured the amount of rainfall for five days. The table below shows the information that the student recorded.

Tip
Think about how you could order the recorded numbers in the table from greatest to least amount of rainfall.

Amount of Rainfall	
Day	**Rainfall (cm)**
Monday	2
Tuesday	4
Wednesday	0
Thursday	3
Friday	1

Describe the findings of this investigation. Include a comparison of the rainfall on the different days.

____

____

____

____

____

____

Name ______________________ Date ______________________

5 You can use tools to take measurements.

> **Tip**
> Remember that mass is the amount of matter in an object.

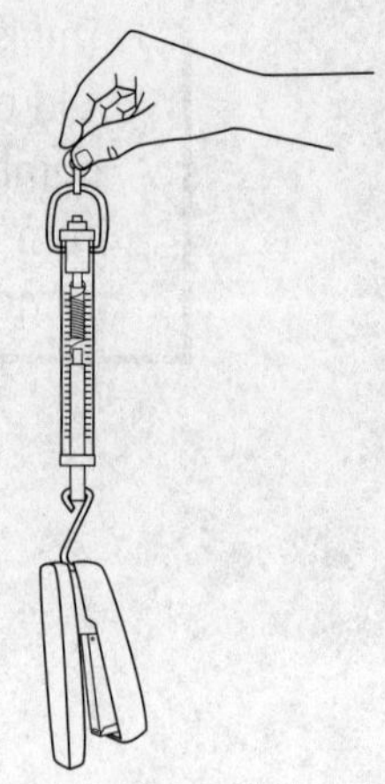

(a) What characteristic of the stapler is the spring scale measuring?

____

____

(b) What tool could you use to measure the stapler's mass? Describe how to use the tool.

____

____

____

____

____

____

____

Name ______________________________

Date ______________________________

Chapter 1 Practice Set

Tip
Think about how a scientist uses each tool.

1 Which tool helps scientists classify organisms they cannot see with their eyes alone?

Ⓐ microscope

Ⓑ thermometer

Ⓒ test tube

Ⓓ meter stick

---

2 Which cell part could be used to classify a living thing as a plant?

Ⓐ vacuole

Ⓑ chloroplast

Ⓒ cell membrane

Ⓓ mitochondria

Name ______________________ Date ______________________

3 The pictures below show several organisms. Study the pictures.

Tip
Recall that invertebrates are animals without a backbone.

  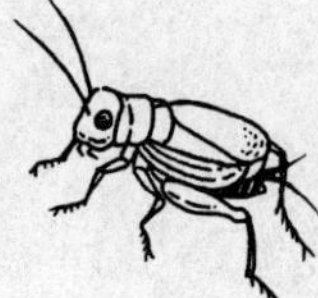 

Scientists can classify each organism into a major group of living things. Which organism is classified as an invertebrate?

Ⓐ the plant

Ⓑ the lizard

Ⓒ the insect

Ⓓ the hawk

Name ______________________ Date ______________________

4 Jacob is making a poster about reptiles. He cut out the four pictures shown below.

> **Tip**
> Think of the characteristics of reptiles.

turtle

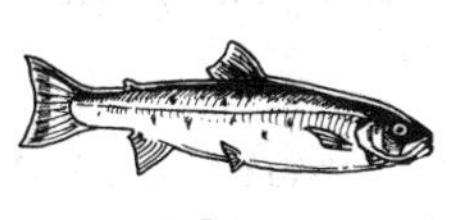

salmon

shark

toad

Which animal should Jacob use for his poster? Explain your answer.

____

____

____

____

____

Name ______________________ Date ______________________

Tip
What characteristics do all arthropods have in common?

5 Manuel and Amanda are hiking along a trail in a forest. Using field glasses, they see an organism walking on a tree branch. The children agree that it is an arthropod. But they are not certain if the arthropod is an insect or an arachnid.

(a) If the arthropod is an insect, what characteristics should the organism have?

____

____

____

(b) If the arthropod is an arachnid, what characteristics should the organism have?

____

____

____

Name ______________________

Date ______________________

1 Which of the following describes the characteristic of eye color?

Ⓐ gene

Ⓑ trait

Ⓒ heredity

Ⓓ nurture

---

2 What happens when a plant is pollinated?

Ⓐ Male sex cells are brought to the female part of the flower.

Ⓑ A pair of leaves forms along all the stems of the plant.

Ⓒ New roots start to grow under the ground.

Ⓓ Seeds travel away from the plant to rich soil.

Tip

The word *pollinate* is related to the word *pollen*.

Name ______________________ Date ______________________

Tip
Think of the stage of life in which people's bodies no longer grow.

3 Miguel has reached his full height and his body has already finished developing. In what stage of development is Miguel?

Ⓐ infancy

Ⓑ childhood

Ⓒ adolescence

Ⓓ adulthood

Name ______________________ Date ______________________

4 The illustration below shows a fern frond. Study the illustration.

> **Tip**
> Remember that ferns reproduce in two different generations.

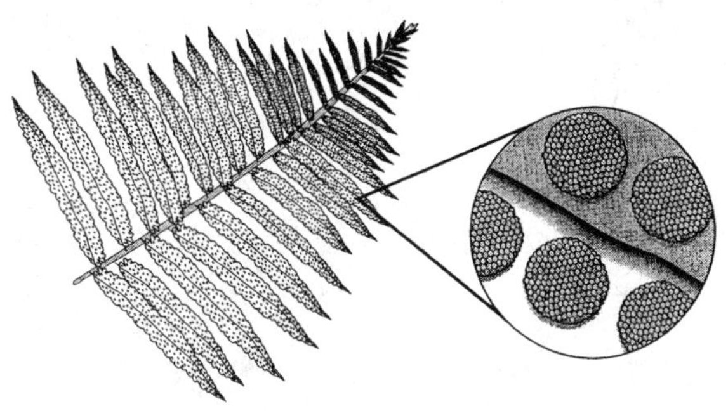

A fern's life cycle is made up of several stages. Describe the stage in a fern's life cycle that is shown in the picture above. Then describe the next stage.

____

____

____

____

____

____

____

Name ______________________ Date ______________________

5 A butterfly undergoes complete metamorphosis.

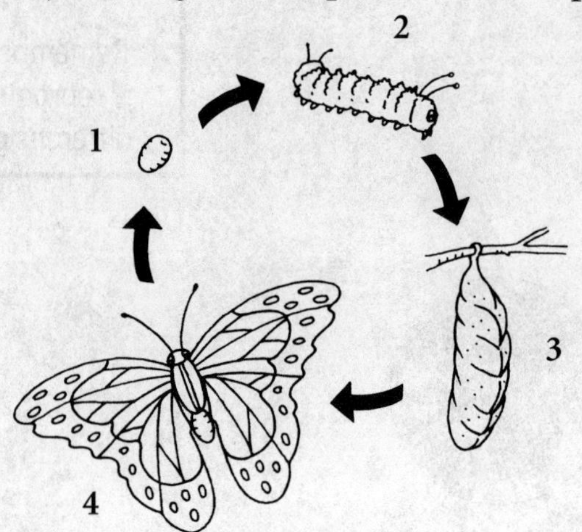

Tip

Recall why leaves are important to plants.

(a) Identify the stages in the butterfly's life cycle shown at each step in the diagram. Describe what happens at each stage.

____

____

____

____

____

____

(b) What might happen to a plant if all the eggs laid by an adult butterfly survive to become larvae?

____

____

____

____

Name ______________________________

Date ______________________________

Tip
Remember that adaptations help organisms survive.

1 Which of the following is an adaptation for getting food?

Ⓐ the sticky tongue of a frog

Ⓑ the flying pattern of a group of birds

Ⓒ the change of fur color in rabbits

Ⓓ hibernation in bears

Name ______________________ Date ______________________

2 Which behavior is based on instinct?

(A) bear cubs looking through a garbage pail for food

(B) birds migrating to the same location every year

(C) lion cubs imitating hunting behaviors of their mother

(D) a dog responding to its owner's commands

---

3 Which organism could be classified as a living fossil?

(A) saber-toothed tiger

(B) Florida camel

(C) ginkgo tree

(D) woolly mammoth

Tip

Think about what the term *living fossil* means. Look for organisms that are still *living* today.

Name ______________________ Date ______________________

4 All living things have basic needs.

> **Tip**
> A beaver's teeth act like a chisel.

Beavers depend on trees to meet the basic needs of food and shelter. How do beavers use parts of trees to meet these two basic needs?

____

____

____

____

____

Name ______________________ Date ______________________

5 Look at the picture of the fossil shown below.

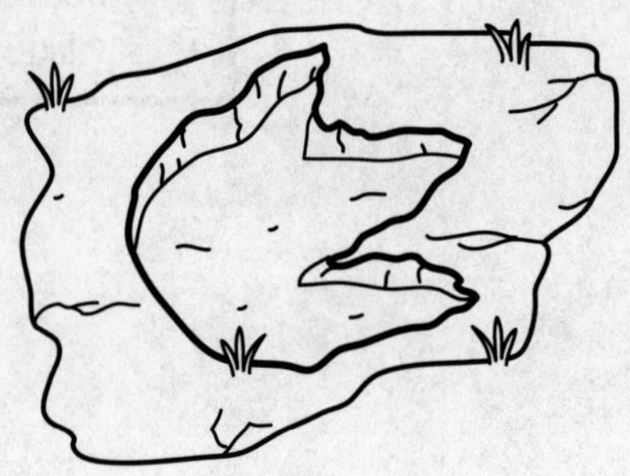

> **Tip**
> Think about what must have happened for the footprint to have become a fossil.

(a) How might the fossil have formed?

____

____

____

____

____

(b) What can be learned from the fossil?

____

____

____

____

____

Name ______________________

Date ______________________

Tip
Think about the function of trees' roots in an ecosystem.

1 A builder uproots trees in order to make room for a new neighborhood. What is the most likely effect of the builder's actions?

Ⓐ Soil will form.

Ⓑ Soil will erode.

Ⓒ Soil will be deposited.

Ⓓ Soil will harden into rock.

Name ______________________ Date ______________________

2 Examine the picture of the forest environment.

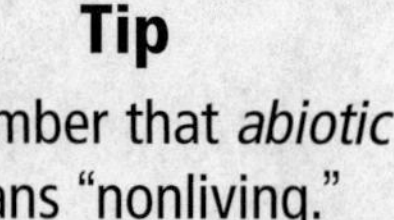
Tip
Remember that *abiotic* means "nonliving."

What are two abiotic factors in the environment shown?

Ⓐ water and rocks

Ⓑ plants and rocks

Ⓒ plants and animals

Ⓓ water and animals

---

3 Which is a way to protect natural resources?

Ⓐ Burn coal for electricity.

Ⓑ Fill in a swamp to build a new playground.

Ⓒ Leave a pond and trees next to a new building.

Ⓓ Grow a thick, green lawn by using fertilizer and weed killer.

Name ______________________ Date ______________________

4 Water hyacinth is a floating plant that reproduces quickly. Mats of water hyacinths can cover a pond or stream. The mats block sunlight to other water plants. They also block land animals from the water.

> **Tip**
> Consider how plants and animals interact with their environment.

What happens to the populations of other plants and animals as water hyacinths grow? Explain why these changes happen.

____

____

____

____

____

____

Name ____________________ Date ____________________

5 The picture on the left shows an ecosystem in an open lot. The picture on the right shows the same place with a new fence.

> **Tip**
> Think of things that do not show in the picture, as well as things that do.

(a) List three biotic and three abiotic factors of the ecosystem.

____

____

____

____

____

(b) What abiotic factor is affected by the fence? How does that change the ecosystem?

____

____

____

____

____

____

Name ______________________________

Date ______________________________

Chapter 5 Practice Set

Tip
Think about which organism would be able to live in a very dry place.

1 Which organism would find its niche in a desert habitat?

Ⓐ an alligator

Ⓑ a crab

Ⓒ an oak tree

Ⓓ a sagebrush

---

2 What is the name for consumers that are eaten by other consumers?

Ⓐ prey

Ⓑ fossils

Ⓒ predators

Ⓓ carnivores

Name ______________________ Date ______________________

3 Which type of organism is at the beginning of every food chain?

Ⓐ decomposer

Ⓑ first-level consumer

Ⓒ producer

Ⓓ top-level consumer

Tip

Organisms that make their own food are at the beginning of every food chain.

Name ______________________ Date ______________________

4 The diagram shows a pond food web.

> **Tip**
> Follow the arrows in the food web.

According to this food web, what are two sources of energy for the bird? Explain your answer.

____

____

____

____

____

Name ______________________ Date ______________________

5 Examine the food chain below involving grass, a prairie dog, and a hawk.

Grass	→	Prairie dog	→	Hawk

Tip

What does an herbivore eat?

(a) Which consumer is an herbivore and which is a carnivore?

Explain.

____

____

____

____

(b) How do all of the living things in the food chain get the energy to live from sunlight?

____

____

____

____

____

Name ______________________________

Date ______________________________

Chapter 6 Practice Set

1 Which property are you observing when you notice how a mineral reflects light?

Ⓐ color

Ⓑ luster

Ⓒ magnetism

Ⓓ shape

---

2 Which of these makes up the largest part of soil?

Tip
Think about how soil forms.

Ⓐ air

Ⓑ humus

Ⓒ water

Ⓓ weathered rock

Name ______________________ Date ______________________

Tip
Recall which type of rock forms in layers.

3 Study the rock below.

Which type of rock is this?

Ⓐ metamorphic

Ⓑ volcanic

Ⓒ igneous

Ⓓ sedimentary

Name ______________________ Date ______________________

4 Sometimes, farmers add nutrients to make soil better for growing plants. The nutrients they add are fertilizers.

Recall what you know about recycling and about fertilizers made in factories.

Identify two categories of fertilizers. Explain how each of these is made.

____

____

____

____

____

Name ______________________ Date ______________________

5 Denzel is writing a report about the rock cycle. He includes this diagram as part of his report.

> **Tip**
> Think about how rocks change from one type to another.

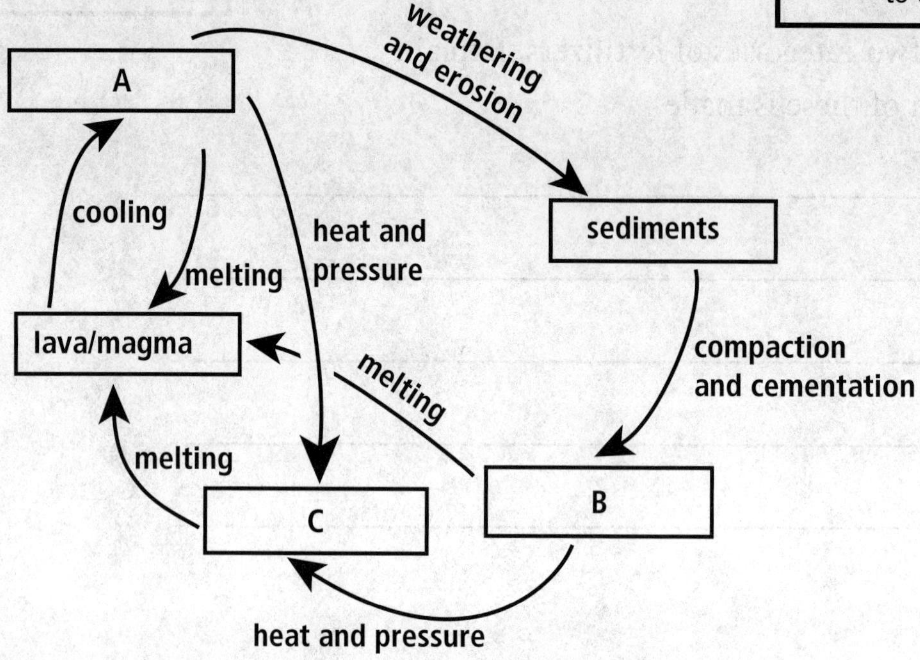

(a) Label the diagram to identify the missing steps.

(b) In his report, Denzel writes that the rock cycle begins with igneous rock, moves on to sedimentary rock, and ends with metamorphic rock. Explain why this statement is false.

____

____

____

____

____

____

Name ______________________________

Date ______________________________

Chapter 7 Practice Set

Tip
Recall that landforms are *natural* features.

1 Which of the following is a landform?

Ⓐ highway

Ⓑ mountain

Ⓒ skyscraper

Ⓓ subdivision

---

2 Which landform occurs at the end of a river?

Ⓐ delta

Ⓑ dune

Ⓒ valley

Ⓓ canyon

Name ______________________ Date ______________________

3 What is one way an organism becomes a fossil?

Ⓐ It is decayed by decomposers.

Ⓑ It reproduces only once.

Ⓒ It is covered by sediment.

Ⓓ It is eaten by other organisms.

Tip

Remember that a fossil provides physical evidence that an organism once existed.

Name ______________________ Date ______________________

4 Look at the geologic time scale below.

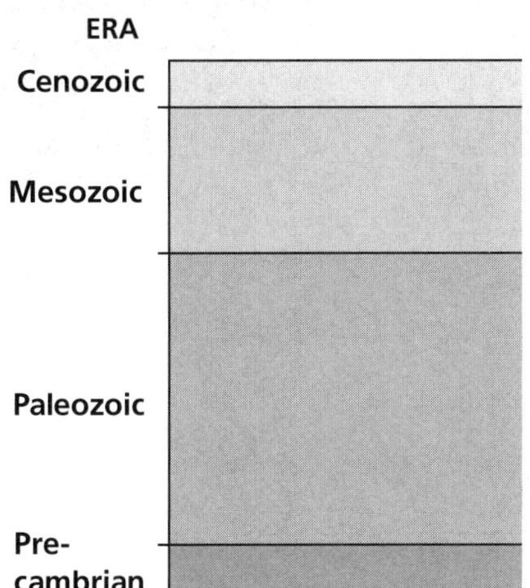

Tip

Try to recall, in order, the names of the geologic eras. Then decide which life forms appeared first and which appeared later.

What kind of life form was dominant on Earth during the Mesozoic era ? What happened to this kind of life form at the end of the era?

____

____

____

____

____

Name ______________________ Date ______________________

Tip

Ask yourself what the outside of a loaf of bread is called. What is the inside of an apple called?

5 The diagram shows Earth's layers. Study the diagram.

(a) Name each layer and describe it.

____

____

____

____

____

____

____

____

(b) Plates in Earth's two upper layers can move. How do two land plates move to form a mountain chain?

____

____

____

____

Name ______________________________

Date ______________________________

Chapter 8 Practice Set

Tip
Recall what you have learned about precipitation.

1 Which forms when water vapor turns directly into ice?

Ⓐ hail

Ⓑ rain

Ⓒ sleet

Ⓓ snow

---

2 Which of the following is a large body of air?

Ⓐ eye

Ⓑ air mass

Ⓒ front

Ⓓ rain shadow

Name ______________________ Date ______________________

3 Look at the diagram of the water cycle.

> **Tip**
> Water is mostly liquid or solid on Earth's surface, but mostly gas in the air.

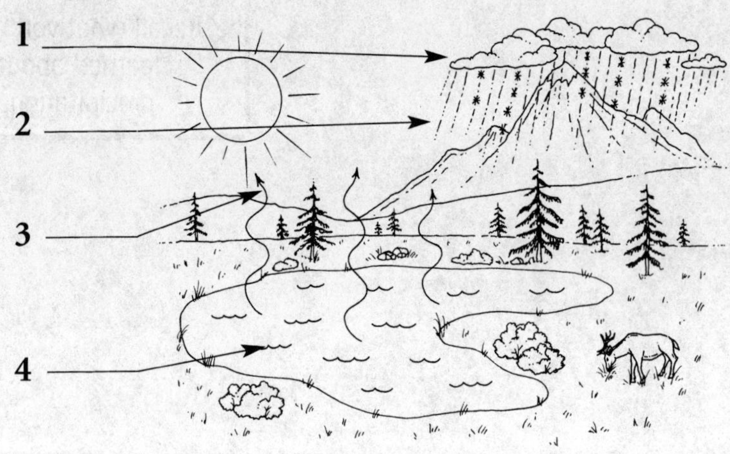

Which point shows the process in which liquid water is changed into a gas?

Ⓐ 1

Ⓑ 2

Ⓒ 3

Ⓓ 4

Name ______________________ Date ______________________

4 A weather front that stays in one place for several days is called a stationary front.

Tip
Think of what happens when two air masses collide.

What kind of weather usually occurs along a stationary front? What does this type of weather often result in?

____

____

____

____

____

____

Name ______________________ Date ______________________

5 Mai's class set up a weather station outside their classroom. It had these three instruments.

> **Tip**
> Think about changes in the way the air feels just before it rains.

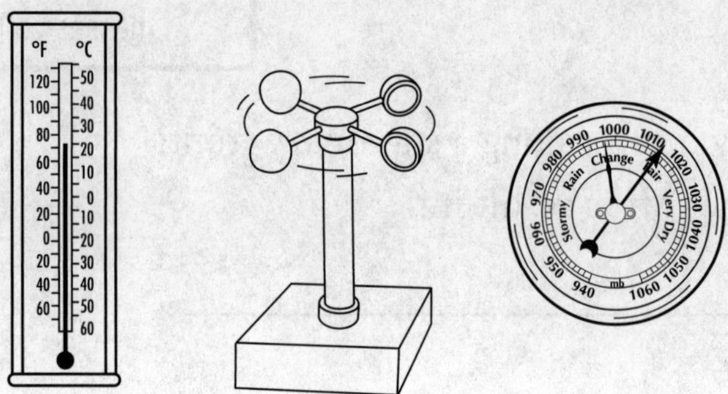

(a) Identify each instrument, and describe what it measures.

____

____

____

____

____

(b) Describe the changes each instrument is likely to show just before it rains. Explain what these changes tell about weather.

____

____

____

____

____

Name ______________________________

Date ______________________________

Chapter 9 Practice Set

Tip
Look for the answer choice in which **both** things cause Earth's seasons.

1 Which of the following causes Earth to have seasons?

Ⓐ Earth's orbit and tilted axis

Ⓑ Earth's tilted axis and size

Ⓒ Earth's orbit and distance from the sun

Ⓓ Earth's orbit and distance from the moon

---

2 Study the diagram of the solar system.

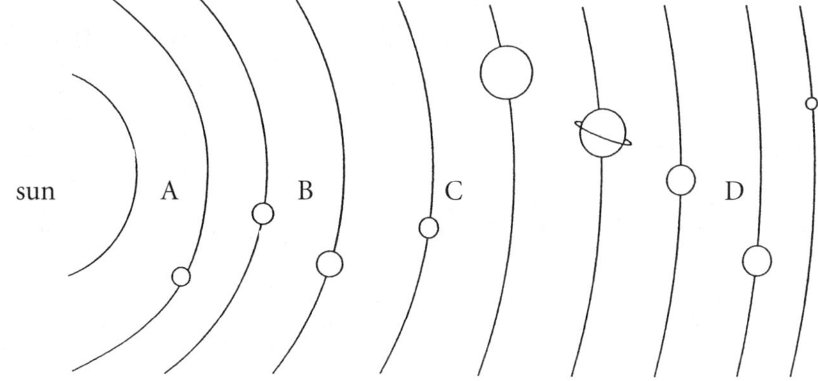

A ring of asteroids separates the inner and outer planets. What point in the diagram below shows where this asteroid belt is located?

Ⓐ Location A

Ⓑ Location B

Ⓒ Location C

Ⓓ Location D

Name ______________________ Date ______________________

3 Look at the diagram of the moon.

> **Tip**
> Cross off two wrong answer choices based on the fact that only part of the moon is lighted.

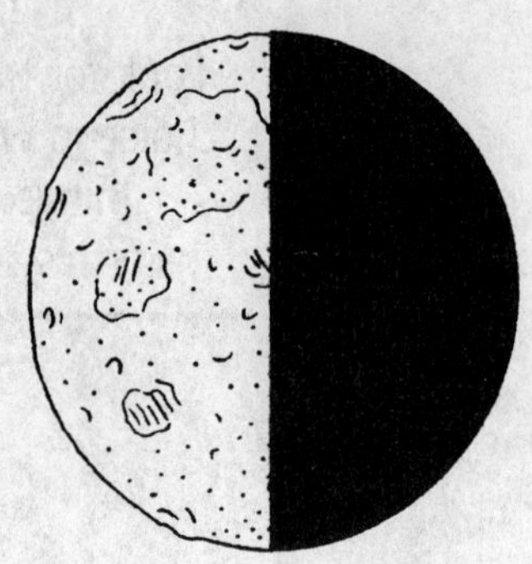

Which moon phase is shown in the diagram?

Ⓐ new moon

Ⓑ first quarter

Ⓒ full moon

Ⓓ third quarter

Name ______________________ Date ______________________

4 Our sun is a medium-sized star. Many other stars are larger than our sun. Yet, when we look into the sky, the sun appears much bigger than any other stars.

> **Tip**
> Think about why an airplane flying in the sky seems to be very small.

Explain why the sun appears to be so large and the other stars appear to be dots of light in the sky.

____

____

____

____

____

Name ______________________ Date ______________________

5 Trina is making a model of the solar system. She knows that a ring of asteroids separates the nine planets into two groups—the inner planets and the outer planets.

Tip

Think of the characteristics of the planets: size, distance from the sun, what they are made of, and number of moons.

(a) Name the inner planets and the outer planets.

____

____

____

____

____

(b) How do the inner planets differ from the outer planets?

____

____

____

____

____

____

____

____

____

____

Name ____________________

Date ____________________

1 Which physical property can you measure with a pan balance?

(A) density

(B) mass

(C) temperature

(D) volume

---

2 Suppose the mass of a block is 30 grams (g) and its volume is 10 cubic centimeters (cm^3). What is the density of the block?

(A) 3 g/cm^3

(B) 30 g/cm^3

(C) 40 g/cm^3

(D) 300 g/cm^3

Tip

Use the measurements given for mass and volume to calculate the density.

Name ______________________ Date ______________________

3 Study the diagram below.

> **Tip**
> Compare the arrangement and motion of particles in each picture.

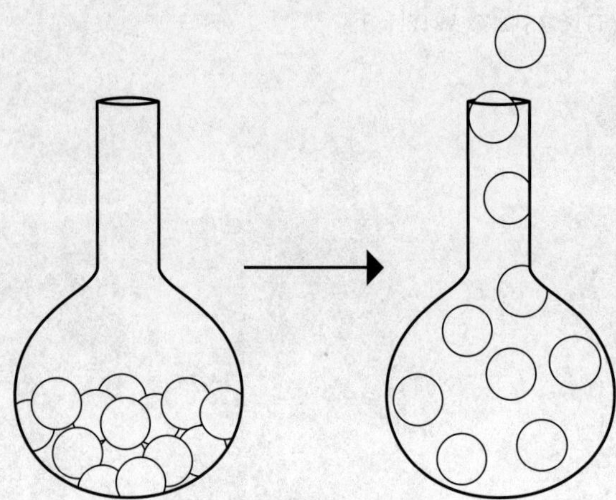

Which description best matches the diagram?

Ⓐ Liquid particles cool, slow down, and freeze into a solid.

Ⓑ Solid particles heat up, move faster, and melt into a liquid.

Ⓒ Liquid particles heat up, move rapidly, and evaporate into a gas.

Ⓓ Gas particles cool, move closer together, and condense into a liquid.

Name ______________________ Date ______________________

4 Look at what Liz ate for lunch today.

Today's Lunch
pepperoni pizza
tossed salad with oil & vinegar dressing
bottled tea with sugar

Tip

Recall the definition of *solution*. Then find the item from the menu that best fits the definition.

Which item from Liz's lunch is a solution? Explain your choice.

____

____

____

____

____

____

____

Name ______________________ Date ______________________

5 A student has a mixture of many beads. Some of the beads have 1-centimeter (cm) diameters, some have 2-cm diameters, and some have 3-cm diameters. The student decides to use filters to separate the beads quickly.

> **Tip**
> The filter holes should be big enough to let small beads pass through. They should be small enough to block large beads.

(a) How many filters does the student need to separate the mixture? What size holes should the filters have?

____

____

____

____

(b) What steps should the student follow to separate the mixture?

____

____

____

____

____

____

Name ______________________________

Date ______________________________

Chapter 11 Practice Set

Tip
Remember that a chemical change produces at least one new substance.

1 Which describes a chemical change?

Ⓐ rust forming on an old wheelbarrow

Ⓑ ice melting into liquid water

Ⓒ paper being shredded

Ⓓ sugar mixing in water

---

2 Which of the following is an element?

Ⓐ air

Ⓑ carbon

Ⓒ steel

Ⓓ water

Name ______________________ Date ______________________

3 Look at the illustration below.

> **Tip**
> Be sure to identify what is happening in the correct order.

Which sequence best explains what is happening in the pictures?

Ⓐ boiling, dissolving

Ⓑ freezing, condensing

Ⓒ dissolving, evaporating

Ⓓ evaporating, condensing

Name ______________________ Date ______________________

4 During science class a student recorded some observations about an unknown metal.

Observations
dull
yellow color
melts at 115°C
reacts with oxygen

Tip
Remember that chemical properties describe how a substance interacts with other substances.

Classify each observation as either a physical property or a chemical property of the substance.

____

____

____

Name ______________________ Date ______________________

5 A physical change and a chemical change occur when a candle burns.

Tip

Remember the signs of physical and chemical changes.

(a) Identify one physical change that occurs as the candle burns.

____

____

(b) Identify one chemical change that occurs as the candle burns. What are some signs of this chemical change?

____

____

____

____

Name ______________________________

Date ______________________________

1 Look at the animals shown below.

Which of the following statements is true?

Ⓐ The lion makes sounds with a higher pitch.

Ⓑ The cat makes sounds with a higher pitch.

Ⓒ The lion and the cat make sounds with the same frequency.

Ⓓ The lion makes sounds with a higher frequency than the cat.

---

2 When sound waves reach your ear, the vibrations are passed from the eardrum, through tiny bones, to the cochlea. How do these vibrations travel from the cochlea to the brain?

Tip
Think about the steps involved in the hearing process.

Ⓐ as echoes

Ⓑ through funnels

Ⓒ through bones

Ⓓ as electrical signals

Name ______________________ Date ______________________

3 Water and sound move in waves. How is the movement of sound waves different from the movement of water waves?

Ⓐ Sound waves cross each other and water waves do not.

Ⓑ Sound waves have longer wavelengths than water waves have.

Ⓒ Sound waves move back and forth, and water waves move up and down.

Ⓓ Sound waves move up and down, and water waves move back and forth.

Tip

Think about what happens as sound waves travel and how waves move through water.

Name ______________________ Date ______________________

4 Adult voices are deeper than children's voices. How do a child's vocal cords differ from an adult's vocal cords? What causes a child's voice to be higher than an adult's voice?

Tip
Think about what causes sounds to be high or low.

____

____

____

____

____

Name ______________________ Date ______________________

5 Notice in the picture below that Jon's room shares a wall with the noisy neighbors next door.

> **Tip**
> Notice that Jon's room has a bare floor and walls.

(a) How do sounds from the neighbor's apartment travel to Jon's ears?

____

____

____

(b) Describe two things Jon could do to his room to absorb more of the neighbor's noises. Explain your reasoning.

____

____

____

____

Name ______________________________

Date ______________________________

Chapter 13 Practice Set

1 Which object is the best conductor of heat?

> **Tip**
> Recall the definition of the term *conductor*.

Ⓐ 

Metal foil

Ⓒ 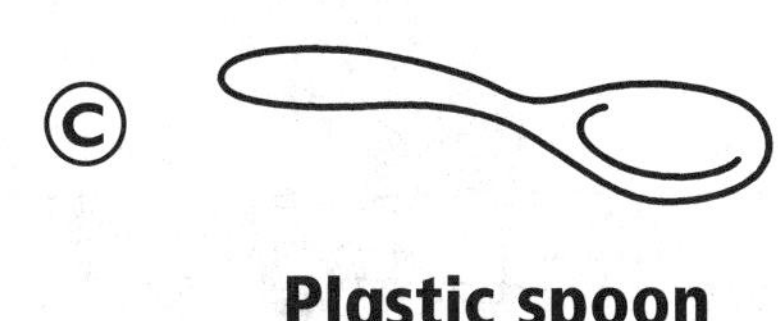

Plastic spoon

Ⓑ **Wooden stick**

Ⓓ **Glass test tube**

---

2 Quincy's mother replaces a light bulb in a lamp. She then puts the lampshade back on the lamp. Quincy cannot see the bulb, but he can see the light through the shade. Which word best describes the lampshade?

Ⓐ opaque

Ⓑ radiant

Ⓒ transparent

Ⓓ translucent

Name ______________________ Date ______________________

3 The pictures below show a transfer of energy.

> **Tip**
> Be sure to match the descriptions to the diagram in the proper order.

Which sequence best describes the pictures?

Ⓐ waste heat, fuel, motion

Ⓑ motion, waste heat, light

Ⓒ fuel, motion, waste heat

Ⓓ waste heat, motion, fuel

Name ______________________ Date ______________________

4 You can see because light enters your eyes. The light can come from an object that gives off light, or the light can bounce off an object and into your eyes.

> **Tip**
> Look at an opaque object, such as your desk. Ask yourself, "Why can I see this object?"

Do opaque objects reflect any light? Explain your reasoning and give an example.

____

____

____

____

Name ______________________ Date ______________________

5 The heater in a car runs off waste heat from the car's engine.

> **Tip**
> Recall the definition of *waste heat*. What can it do?

(a) Carol's dad keeps the inside of their car cold, even on cool days. He says that not using the car's heater helps save gasoline. Is Carol's dad correct? Explain your answer.

____

____

____

____

(b) Carol is cold when her dad drives her to school. Write something she could say to persuade her dad to turn on the car's heater. Include a definition of *waste heat* and an explanation of what it can and cannot do.

____

____

____

____

____

____

Name ______________________________

Date ______________________________

Tip
The key word is *path*.

1 What is the name of the path that an electric current follows?

Ⓐ circuit

Ⓑ electromagnet

Ⓒ motor

Ⓓ switch

---

2 The picture below shows a way of making electricity by using the energy of falling water.

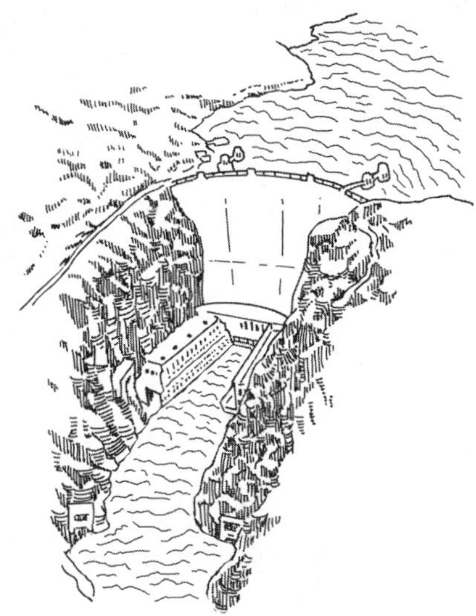

What is this source of electricity called?

Ⓐ chemical power

Ⓑ geothermal power

Ⓒ hydroelectric power

Ⓓ solar power

Name ______________________ Date ______________________

3 Study the diagram of the parallel circuit. Notice that both bulbs are lit.

> **Tip**
> Recall how current flows in a parallel circuit.

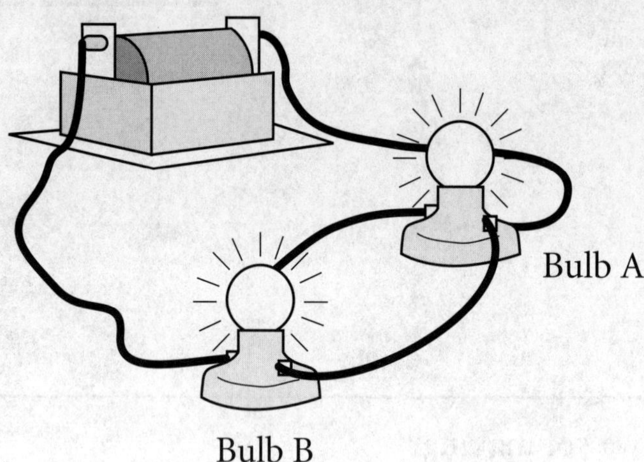

What will happen if you remove one of the wires between Bulb A and Bulb B?

Ⓐ The light of both lamps will be dimmer.

Ⓑ The light of both lamps will be brighter.

Ⓒ Current will no longer flow through the circuit.

Ⓓ The current and lamp brightness will not change.

Name ______________________ Date ______________________

4 Stephen has the objects shown below.

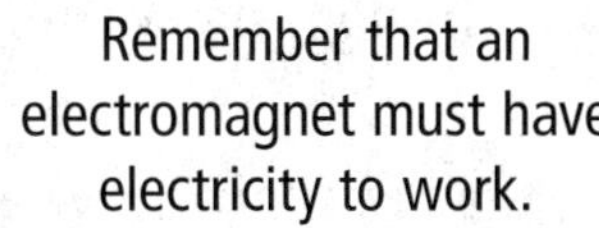

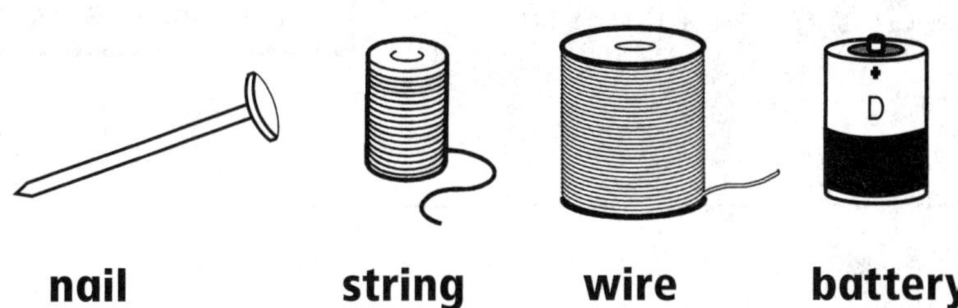

Which objects should Stephen use to make an electromagnet? How can he make an electromagnet with those objects?

____

____

____

____

____

Name ______________________ Date ______________________

5 Electrical energy and chemical energy can change into other forms of energy.

Tip
Think about what you observe when you turn on a lamp or a toy car.

(a) Explain how energy is changed in a lamp that is plugged into a wall socket.

____

____

(b) A toy car uses a battery to run. Explain how the stored energy changes into another form of energy when the toy car is turned on.

____

____

____

Name ______________________________

Date ______________________________

1 Which of the following objects is accelerating?

Ⓐ A car slows down at a yield sign.

Ⓑ A train travels west at a constant speed.

Ⓒ A truck travels on a highway at the speed limit.

Ⓓ A tractor is parked in a farmer's field.

---

2 Study the table below.

Ball	Diameter (mm)	Mass (grams)
Baseball	73	145
Basketball	240	612
Beach ball	315	48
Softball	95	170

Tip
Remember that inertia is a measure of how hard it is to change an object's speed.

Which ball has the greatest inertia?

Ⓐ baseball

Ⓑ basketball

Ⓒ softball

Ⓓ tennis ball

Name ______________________ Date ______________________

3 Look at the picture of the tree and acorns below.

> **Tip**
> Remember what the "pull force" of Earth is called.

Which force causes the acorns to move toward Earth?

Ⓐ chemical force

Ⓑ electrical force

Ⓒ gravitational force

Ⓓ magnetic force

Name ______________________ Date ______________________

4 Look at the pictures of the children shown below.

> **Tip**
> Think about how you would describe to a friend the motion shown in each picture.

Which children are accelerating? Explain your answer.

____

____

____

____

Name ______________________ Date ______________________

5 Myra ran in a race. The table below shows information about the race.

> **Tip**
> Remember the equation for speed.

Race Information	
length	10 kilometers
temperature	20°C
number of runners	650
Myra's time	1 hour

(a) Use the information in the table to write a number sentence you can use to find Myra's speed. Then calculate Myra's speed.

____

____

(b) Think about the difference in speed and velocity. What additional information would you need to find Myra's velocity?

____

____

Name ______________________________

Date ______________________________

Chapter 16 Practice Set

Tip
Consider whether the wheel acts as a pivot point.

1 Mr. Simmons is using a wheelbarrow to do work. Look at the picture.

A

Which term identifies the point that is labeled A?

Ⓐ bar

Ⓑ force

Ⓒ fulcrum

Ⓓ lever

---

2 Look at the seesaw in the picture below.

Which type of simple machine is the seesaw?

Ⓐ lever

Ⓑ pulley

Ⓒ inclined plane

Ⓓ wheel-and-axle

Name ______________________ Date ______________________

3 Study the picture of the single, fixed pulley shown below.

Tip
Think about how a single, fixed pulley changes the way a job is done.

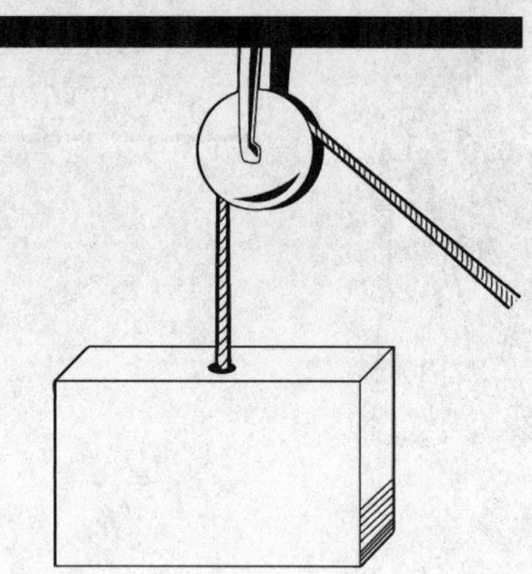

How does the pulley work?

Ⓐ The pulley makes the box lighter and easier to use.

Ⓑ The pulley changes the direction of the force used to lift the box.

Ⓒ The pulley changes the amount of force needed to lift the box.

Ⓓ The pulley takes away friction, which makes the box easier to lift.

Name ______________________ Date ______________________

4 You have been asked to make a poster showing pictures of machines that use a wheel-and-axle. The first two examples you think of are a skateboard and a doorknob. Study the pictures below.

> **Tip**
> Recall how a wheel-and-axle works.

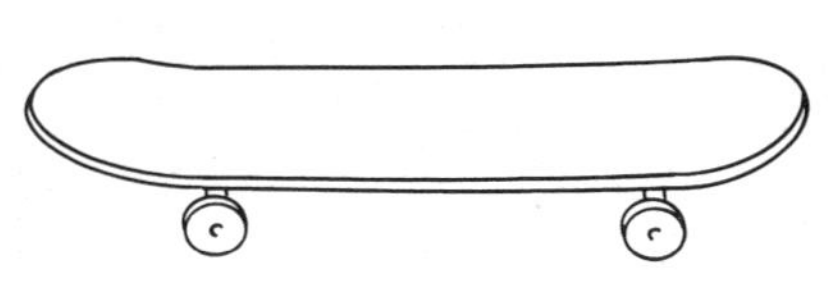

Does either the skateboard or the doorknob use a wheel-and-axle? Explain why or why not.

____

____

____

____

____

Name ____________________ Date ____________________

5 Cora will investigate how the slope of an inclined plane affects the ease with which work is done. She plans to use three wooden ramps that vary only in steepness. She will attach a string to a block of wood and slowly pull the block up each ramp one time. Instead of using a tool to measure force, she will use her own judgment to decide how difficult it is to pull the block up each ramp.

> **Tip**
> Identify things that are the same for each ramp.

(a) Describe one part of Cora's plan that is well planned.

____

____

____

____

____

(b) Suggest one way Cora could improve her investigation.

____

____

____

____

Name ______________________

Date ______________________

1 Shelly pours 100 mL of three different liquids into separate beakers. She heats the beakers on a hot plate and watches them until each begins to boil. Which question can Shelly answer from her observations?

Ⓐ What is the boiling point of each liquid?

Ⓑ How long does it take each liquid to begin boiling?

Ⓒ Which liquid takes the longest to begin boiling?

Ⓓ What was the final temperature of each liquid?

Standard 1, S1.1b

2 A student made a model of a ramp used to load boxes onto a truck. The model is pictured below. The student wants to use the model to find ways to make a real ramp safer.

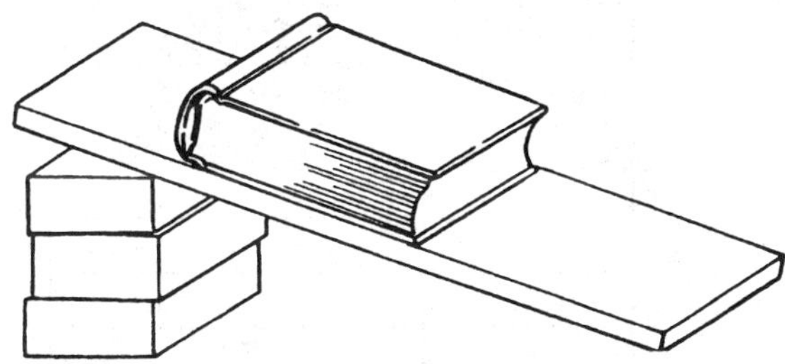

How could the student change the ramp to make it safer?

Ⓐ He could use a different book.

Ⓑ He could make the surface rougher.

Ⓒ He could change the ramp's location.

Ⓓ He could increase the height of the ramp.

Standard 1, T1.1c

Name ______________________________

Date ______________________________

Practice Test 1

3 Look at the objects on the balance below.

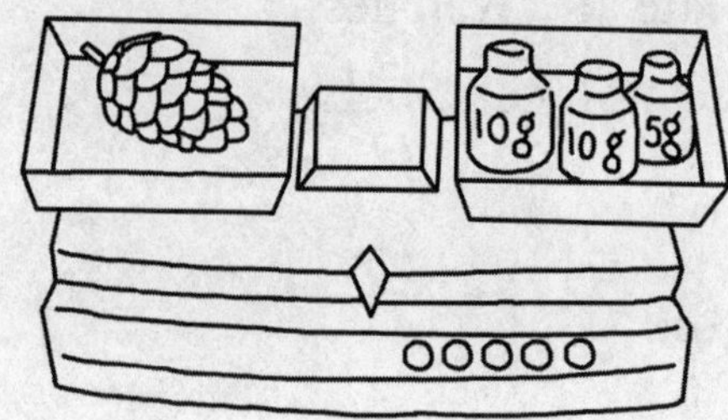

What is the mass of the pinecone?

Ⓐ 5 g

Ⓑ 15 g

Ⓒ 20 g

Ⓓ 25 g

Standard 1, M3.1a

---

4 A student tested the strength of electromagnets. She wanted to see how many paper clips each electromagnet could pick up. Her data table is shown here.

Strength of Electromagnets

Number of Coils	Number of Paper Clips Picked Up
70	8
100	14
150	20
200	27

What can you conclude about an electromagnet's strength from this data?

Ⓐ Strength decreases as the number of coils increases.

Ⓑ Strength increases as the number of coils increases.

Ⓒ The strength is equal to half the number of coils.

Ⓓ The strength is not related to the number of coils.

Standard 1, S3.2a

5 Corn seedlings grow very quickly in the first five days after germination. Study the information in the table below.

Day	Amount of Growth per Day in Centimeters
1	1
2	2
3	4
4	8
5	?

Based on the data in the table, how many centimeters will the corn seedlings most likely grow on Day 5?

Ⓐ 8 centimeters

Ⓑ 16 centimeters

Ⓒ 32 centimeters

Ⓓ 64 centimeters

Standard 1, M1.1c

---

6 Daniel searches the Internet to learn about polar bears. He reads the following statements. Which statement is an opinion instead of a fact?

Ⓐ A polar bear's white fur hides it in the winter.

Ⓑ Polar bears are very good swimmers.

Ⓒ The fur on a polar bear's feet protect it from the cold.

Ⓓ The polar bear is the most beautiful of all bears.

Standard 2, Key Idea 3

Name ______________________

Date ______________________

7 What materials make up most soils?

Ⓐ clay and plaster

Ⓑ sand and cement

Ⓒ pebbles and dead leaves

Ⓓ weathered rock and humus

Standard 4: Physical Setting, 2.1d

8 Humans organize time into units based on the natural motions of Earth. Which unit of time is based on the moon's orbit?

Ⓐ hour

Ⓑ minute

Ⓒ month

Ⓓ year

Standard 4: Physical Setting, 2.1d

9 The constellation Orion can be seen in the night sky in winter. A few months later, it may no longer be visible. What causes this to occur?

Ⓐ Earth's revolution around the sun

Ⓑ Earth's rotation on its axis

Ⓒ the moon's revolution around Earth

Ⓓ Orion's revolution around the sun

Standard 4: Physical Setting, 1.1c

Name ___________________________

Date ___________________________

10 What is the best definition for the term *weather*?

Ⓐ average temperature in an area over a long period of time

Ⓑ condition of the outside air at a particular moment

Ⓒ movements in the lower portion of the atmosphere

Ⓓ process of water moving through the air from place to place

Standard 4: Physical Setting, 2.1a

---

11 Why do many electrical wires have a plastic coating?

Ⓐ Plastic is a good electrical insulator.

Ⓑ Plastic forms a circuit with the wire.

Ⓒ Plastic is a good conductor of electricity.

Ⓓ Plastic is a switch, which opens the circuit.

Standard 4: Physical Setting, 4.1c

---

12 Your body takes in energy from the food you eat. What type of energy change occurs as your body uses this stored energy to walk or run?

Ⓐ electrical energy to chemical energy

Ⓑ chemical energy to mechanical energy

Ⓒ mechanical energy to electrical energy

Ⓓ heat energy to chemical energy

Standard 4: Physical Setting, 4.2a

Name ______________________

Date ______________________

13 Think about the attraction of the two magnets shown here.

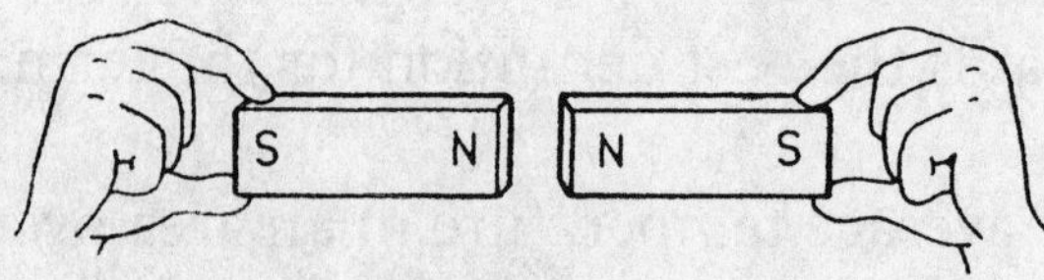

What would happen if you placed some water or a wooden board between the magnets?

Ⓐ The magnets would attract through both the board and the water.

Ⓑ The magnets would attract through the board but not through the water.

Ⓒ The magnets would attract through the water but not through the board.

Ⓓ The magnets would not attract through either the board or through the water.

Standard 4: Physical Setting, 5.2a

---

14 This table shows the freezing points and boiling points for different types of metal.

Metal	Melting Point (°C)	Boiling Point (°C)
Cobalt	1495	2928
Iridium	2447	4428
Nickel	1455	2914
Silver	1541	2836

A technician places a cube of each metal into an oven. As the temperature in the oven rises, which cube will change to a liquid first?

Ⓐ cobalt

Ⓑ iridium

Ⓒ nickel

Ⓓ silver

Standard 4: Physical Setting, 3.2b

Name ______________________________

Date ______________________________

15 The pictures below show stages in a sphinx moth's life cycle. The pictures are not in the proper order.

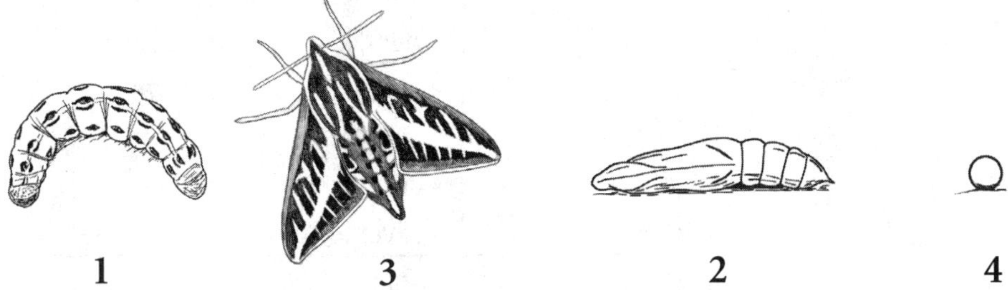

Which answer shows the correct order of the stages?

Ⓐ 1 → 2 → 3 → 4

Ⓑ 3 → 2 → 1 → 4

Ⓒ 4 → 1 → 2 → 3

Ⓓ 4 → 3 → 2 → 1

Standard 4: Living Environment, 4.1e

---

16 Which reason best explains why the decay of an organism is an important part of its life cycle?

Ⓐ Decay keeps the environment clean.

Ⓑ Decay returns nutrients to the soil.

Ⓒ Decay provides food for predators.

Ⓓ Decay provides a new habitat for bacteria.

Standard 4: Living Environment, 6.1d

Name ______________________________

Date ______________________________

17 This table shows some physical properties of substances.

Substance	Physical Properties
water	• colorless • odorless • liquid at room temperature
silver	• shiny • soft • silver in color
iron	• shiny • hard • grayish-silver in color
sulfur	• dull • brittle • yellow

Which two senses would be most helpful in identifying the substances?

Ⓐ sight and smell

Ⓑ sight and touch

Ⓒ sight and taste

Ⓓ smell and touch

Standard 4: Living Environment, 5.2c

---

18 What happens to a population when its food supply becomes scarce?

Ⓐ It stays the same.

Ⓑ It decreases in number.

Ⓒ It increases in number.

Ⓓ It dies out completely.

Standard 4: Living Environment, 6.1f

Name ____________________

Date ____________________

19 The graph below shows the heart rate of bears during different months.

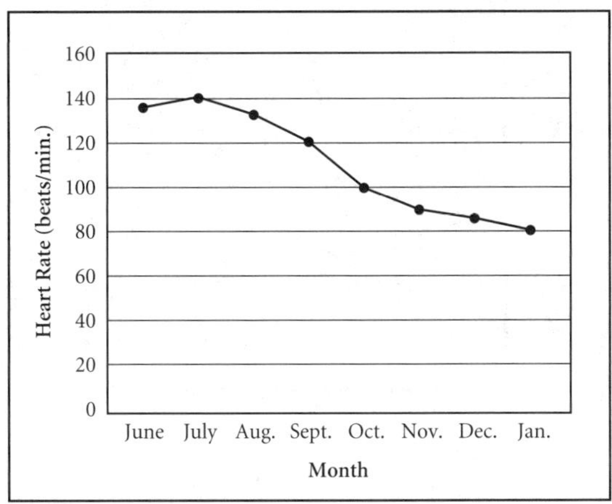

Which adaptation is represented by the graph?

Ⓐ extinction

Ⓑ migration

Ⓒ hibernation

Ⓓ learned behaviors

Standard 4: Living Environment, 5.2f

---

20 Chloroplasts in the leaves of plants use sunlight, carbon dioxide, and water to make sugar, a food. What is this process called?

Ⓐ reproduction

Ⓑ photosynthesis

Ⓒ respiration

Ⓓ digestion

Standard 4: Living Environment, 6.2a

Name ______________________________

Date ______________________________

21 This diagram shows part of an aquatic food web.

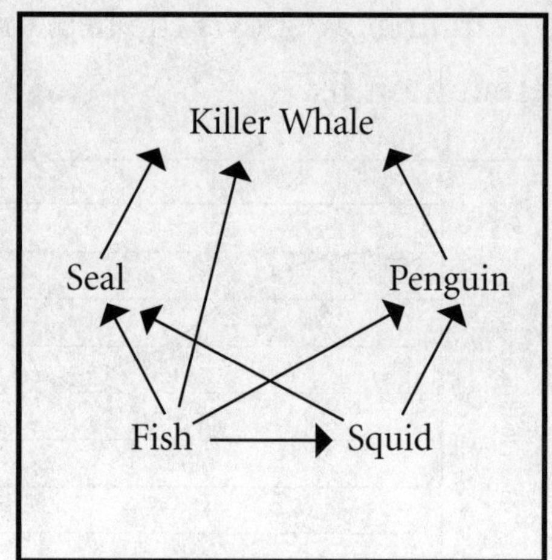

The diagram shows the transfer of energy among the organisms. From which animal does the squid get its energy?

Ⓐ seal

Ⓑ fish

Ⓒ killer whale

Ⓓ penguin

Standard 4: Living Environment, 6.1b

---

22 The Incas lived in the Andes Mountains, which have steep slopes. To have more land to grow crops, they cut terraces, or flat areas, into the slopes. How would building terraces to grow food have affected the Inca population?

Ⓐ Building terraces would have taken away areas where people built their homes.

Ⓑ The terraces would have allowed more food to be grown, so the population would have increased.

Ⓒ Building the terraces would have caused erosion on the steep hillsides, resulting in dangerous floods.

Ⓓ The terraces would have provided homes for predators that might injure people.

Standard 4: Living Environment, 7.1a

Name ______________________________

Date ______________________________

23 Many scientists think the planet is getting warmer. Scientists hypothesize that this increase in temperature is caused by an increase in carbon dioxide in the air. Which of the following is the most likely cause of the increase in carbon dioxide in the air?

Ⓐ animals breathing

Ⓑ volcanoes erupting

Ⓒ plants going extinct

Ⓓ cars giving off exhaust

Standard 4: Living Environment, 7.1b

24 Which of the following is most responsible for a plant obtaining the water it needs to survive?

Ⓐ leaves

Ⓑ petals

Ⓒ roots

Ⓓ stem

Standard 6, Key Idea 1

Name ______________________________

Date ______________________________

25 This table shows the height of a tree each year for 5 years.

Year	1	2	3	4	5
Height (meters)	4.5	6.5	8	9	9

Why did the tree most likely stop growing after the fourth year?

Ⓐ It had grown as high as it could grow.

Ⓑ It had died after the fourth year.

Ⓒ It did not receive enough sunlight.

Ⓓ It did not receive enough water.

Standard 6, Key Idea 3

26 Krystal is trying to find out which kind of bike is fastest. She tested three different bikes, using the same rider each time. She recorded the speed of each bike after one ride. Her data is shown in this table.

Bike	Speed (km/hour)
Bike A	19
Bike B	12
Bike C	16

How could Krystle obtain more reliable data for this experiment?

Ⓐ She could use three different riders for the bikes.

Ⓑ She could use bikes that have different-sized wheels.

Ⓒ She could test each of the bikes on different types of roads.

Ⓓ She could include more trips and find each bike's average speed.

Standard 7, Key Idea 2

Name ______________________________

Date ______________________________

27 Roberto noticed a lot of erosion on one hillside following a heavy rain. He wonders why one hillside shows more erosion than others. He sets up the experiment shown below. Using a watering can to simulate rain, Roberto pours the same amount of water at the same speed on each tray.

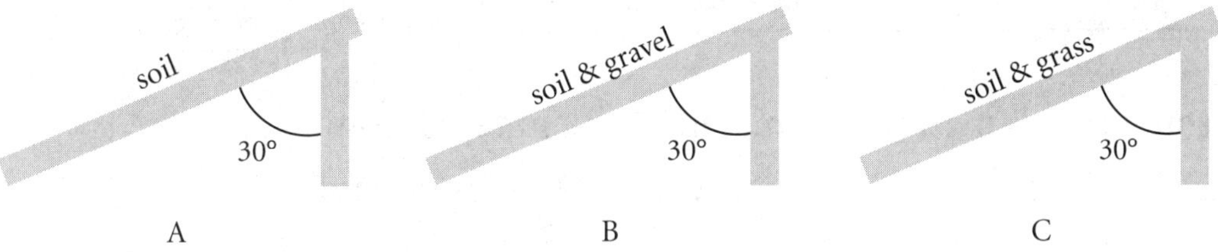

Identify the controlled variables and the variable that is tested in Roberto's experiment.

____

____

____

____

Standard 1, S2.3a

Name ______________________________

Date ______________________________

28 Look at the list of items below.

iron nail	plastic ruler
wooden toothpick	sheet of paper
copper block	steel paper clip
rubber ball	aluminum foil

If you brought each item near a magnet, which items would the magnet attract, and which items would the magnet not attract?

____

____

____

____

____

Standard 4: Physical Setting, 3.1f

Name ______________________________

Date ______________________________

29 A thermometer in a weather station measured the high and low temperatures in Chicago, Illinois, for five days. The temperatures were recorded in the table shown here.

Day	High Temperature	Low Temperature
1	32°F	4°F
2	40°F	12°F
3	38°F	10°F
4	12°F	2°F
5	8°F	1°F

What season of the year do you think it was? Explain your answer.

____

____

____

____

Standard 4: Physical Setting, 1.1a

Name ______________________________

Date ______________________________

30 Examine the weather instrument below.

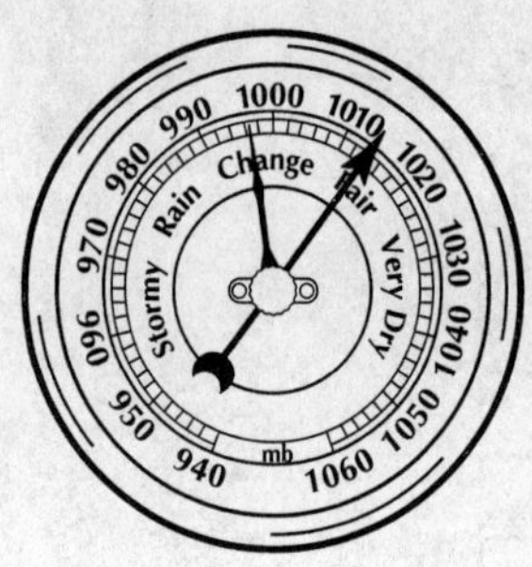

What is this instrument and how can it be used to predict weather?

____

____

____

____

____

____

____

____

Standard 4: Physical Setting, 2.1b

Name ____________________

Date ____________________

31 Think about the forces acting on the bicycle shown in the picture below.

(a) How can unbalanced forces cause the boy's direction of motion to change?

____

____

____

____

(b) How can unbalanced forces cause the boy's direction of motion to change?

____

____

____

____

Standard 4: Physical Setting, 5.1b

Name ______________________________

Date ______________________________

32 A snowshoe hare's fur is rusty-brown during the summer. The fur turns white in winter.

(a) Explain how the change of fur color helps keep the hare safe during the seasons.

____

____

____

____

(b) Identify two other adaptations that might protect an animal that lives in a cold environment.

____

____

____

____

Standard 4: Living Environment, 5.2e

Name ______________________________

Date ______________________________

1 Nancy tests how many paper clips a room-temperature magnet will pick up. She then warms the magnet by leaving it in the sun for an hour. Afterwards, she tests how many paper clips the warm magnet will pick up. What is a possible hypothesis for Nancy's experiment?

Ⓐ How many paper clips will the warm magnet pick up?

Ⓑ The magnet will pick up more paper clips if it is warm.

Ⓒ The warm magnet did not pick up more paper clips.

Ⓓ How can I make the magnet pick up more paper clips?

Standard 1, S1.3a

2 Ken pours water into three containers. He measures the amount of water in each container three days later. This table shows his measurements.

Container	Water on Day 1	Water on Day 4
small jar with lid	150 mL	150 mL
small jar without lid	150 mL	120 mL
large bowl	150 mL	75 mL

What can Ken conclude from the measurements?

Ⓐ The larger the surface area of water exposed to air, the faster the water will evaporate.

Ⓑ The larger the surface area of water exposed to air, the slower the water will evaporate.

Ⓒ The larger the surface area of water that is covered, the faster the water will evaporate.

Ⓓ The larger the surface area of water that is covered, the slower the water will evaporate.

Standard 1, M2.1a

Name ______________________________

Date ______________________________

Practice Test 2

3 Rosa wants to find out which conditions are best for growing beans. She designs the following experiment.

Setup A: flowerpot with 50 grams of soil, 5 bean seedlings, 8 hours of light each day, 100 milliliters of water each day

Setup B: identical flowerpot with 50 grams of same soil, 5 bean seedlings, 10 hours of light each day, 200 milliliters of water each day

Setup C: identical flowerpot with 50 grams of same soil, 5 bean seedlings, 10 hours of light each day, 100 milliliters of water each day

Another student points out that the experiment has too many variables. What is one way Rosa could change the experiment to avoid this problem?

Ⓐ She could leave the beans in the sun for fewer hours.

Ⓑ She could use a different type of soil in each pot.

Ⓒ She could have twice as many seedlings in each setup.

Ⓓ She could give each seedling the same amount of water.

Standard 1, S2.2a

---

4 Kim is making a model of the solar system. She wants the farthest planet to be 1 meter away from the sun in her model. How can she determine how far from the sun to place the Earth?

Ⓐ add Earth's real distance and Pluto's real distance

Ⓑ subtract Earth's real distance from Pluto's real distance

Ⓒ multiply Earth's real distance by Pluto's real distance

Ⓓ divide Earth's real distance by Pluto's real distance

Standard 1, T1.3a

Name ______________________

Date ______________________

5 Tara measures the speed at which four toy cars roll down a ramp. She predicts that longer cars will have a slower speed. The table shows her data.

Car	Length (mm)	Volume (cm3)	Mass (g)	Speed (m/s)
1	8.0	26.5	23.0	0.8
2	6.0	18.0	35.0	1.2
3	7.5	22.4	25.5	1.0
4	7.7	21.5	22.4	0.7

Based on the data in the table, how should Tara modify her prediction?

Ⓐ The longer cars will have a faster speed.

Ⓑ The cars with greater volume will have a slower speed.

Ⓒ The cars with lower masses will have a slower speed.

Ⓓ The shorter cars will have a faster speed.

Standard 1, S3.4a

---

6 Some students measured the water they needed for an experiment. The table shows the volume of water each student started with and the amount of water each student used.

Student	Beginning Volume (mL)	Amount Used (mL)
Allison	75	53
Ben	110	86
Carlos	80	52
Dani	45	22

Which of the following shows how you can determine which student had the most water left at the end of the experiment?

Ⓐ beginning volume + amount used

Ⓑ beginning volume × amount used

Ⓒ beginning volume – amount used

Ⓓ beginning volume ÷ amount used

Standard 1, M1.1b

Name ______________________________

Date ______________________________

7 Which resource would be most useful for writing a description of your observations during an experiment?

Ⓐ Internet encyclopedia

Ⓑ database program

Ⓒ graphing software

Ⓓ word processor

Standard 2, Key Idea 1

8 The diagram below shows the water cycle.

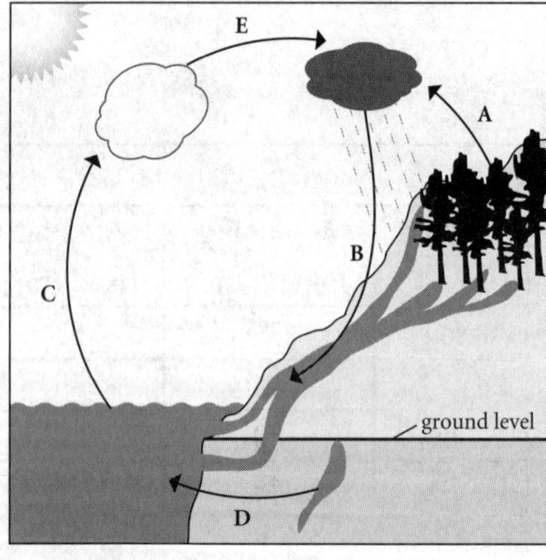

What occurs at point B in the water cycle?

Ⓐ condensation

Ⓑ evaporation

Ⓒ groundwater

Ⓓ precipitation

Standard 4: Physical Setting, 2.1c

Name ______________________________

Date ______________________________

9 Examine the weather instrument shown below.

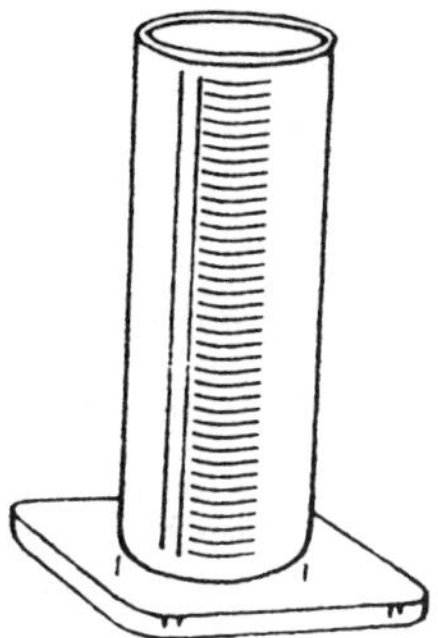

What kind of data can you collect with this instrument?

Ⓐ wind speed

Ⓑ air pressure

Ⓒ air temperature

Ⓓ amount of precipitation

Standard 4: Physical Setting, 2.1b

---

10 In a river valley, which causes the greatest amount of erosion?

Ⓐ high temperatures

Ⓑ bright light

Ⓒ moving water

Ⓓ blowing wind

Standard 4: Physical Setting, 2.1d

Name ______________________

Date ______________________

11 If you watched the stars in the sky all night long, they would appear to move in a large circle. What causes this apparent movement?

Ⓐ Earth's rotation

Ⓑ the moon's orbit

Ⓒ the galaxy's spin

Ⓓ the sun's revolution

Standard 4: Physical Setting, 1.1c

---

12 You throw a baseball to a friend. Which of the following is the best description of the baseball's path as it moves toward your friend?

Ⓐ on a path straight to your friend

Ⓑ on a path curved downward

Ⓒ in the same direction as the throw

Ⓓ in the opposite direction as the throw

Standard 4: Physical Setting, 5.1c

Name ______________________________

Date ______________________________

13 Jenny measured the following properties of a block of metal. Which measurement could she find using only a balance?

Ⓐ volume = 7.6 cm^3

Ⓑ mass = 68 g

Ⓒ length = 2.0 cm

Ⓓ density = 8.96 g/cm^3

Standard 4: Physical Setting, 3.1e

---

14 A girl made the ramp shown below out of a board and a wooden block. She wants to investigate how fast a marble will roll down the ramp. What can she do to slow the marble as it moves down the ramp?

Ⓐ Rub a thin layer of hand lotion into the ramp.

Ⓑ Glue thick carpet onto the ramp.

Ⓒ Sand the ramp to remove rough spots.

Ⓓ Pour several ounces of cooking oil on the ramp.

Standard 4: Physical Setting, 5.1d

Name ______________________________

Date ______________________________

15 Which of these names the process that results in parents and offspring having similar traits?

Ⓐ heredity

Ⓑ survival

Ⓒ reproduction

Ⓓ nurture

Standard 4: Living Environment, 2.1b

16 Look at the webbed feet of the platypus.

What do webbed feet help the platypus do?

Ⓐ fly

Ⓑ swim

Ⓒ walk

Ⓓ catch food

Standard 4: Living Environment, 3.1a

Name ______________________________

Date ______________________________

17 When seasons change, some animals move to another place. What is this behavior called?

Ⓐ hibernation

Ⓑ hiding

Ⓒ migration

Ⓓ hunting

Standard 4: Living Environment, 3.1c

---

18 Ferns grow from spores. From what do flowering plants grow?

Ⓐ seeds

Ⓑ pollen

Ⓒ fiddleheads

Ⓓ cones

Standard 4: Living Environment, 4.1b

---

19 The Arctic fox lives in the Arctic regions of North America. In summer, the fox's fur is reddish-brown. In winter, its fur turns white. How does this color change help the fox survive?

Ⓐ Its color helps the fox blend in with its surroundings.

Ⓑ White fur keeps the fox warmer during the winter.

Ⓒ Reddish-brown fur absorbs heat during the summer.

Ⓓ Its color changes the warmth of the fur from season to season.

Standard 4: Living Environment, 5.1b

Name ______________________________

Date ______________________________

20 Which of the following is a producer in an ecosystem?

Ⓐ green plants

Ⓑ lizards

Ⓒ birds

Ⓓ fish

Standard 4: Living Environment, 6.1a

---

21 Suppose a new kind of plant is introduced into an ecosystem. The new plant spreads rapidly and crowds out existing plants. How will the ecosystem most likely change?

Ⓐ The new plants will poison the soil as they die.

Ⓑ The ecosystem's animal populations will grow.

Ⓒ The populations of existing plants will decrease.

Ⓓ The new plants will pollute the ecosystem's water.

Standard 4: Living Environment, 5.2g

---

22 A snake has a sensory organ that can sense its prey's body heat. How would the snake behave if it sensed body heat to its left?

Ⓐ The snake would stop moving.

Ⓑ The snake would move to its left.

Ⓒ The snake would move to its right.

Ⓓ The snake would ignore the warm body.

Standard 4: Living Environment, 5.2b

Name ______________________________

Date ______________________________

23 Study the living things shown below.

What is the original source of energy for these living things?

Ⓐ grass

Ⓑ the sun

Ⓒ a river

Ⓓ trees

Standard 4: Living Environment, 6.2b

---

24 A worker needs to move a heavy box onto a platform that is several meters above the ground. She wants to lift the box instead of pushing it. Which simple machine should she use?

Ⓐ lever

Ⓑ wedge

Ⓒ pulley

Ⓓ inclined plane

Standard 6, Key Idea 6

Name ______________________________

Date ______________________________

25 This line graph shows the cost of a liter of propane, a type of fuel, since 1996.

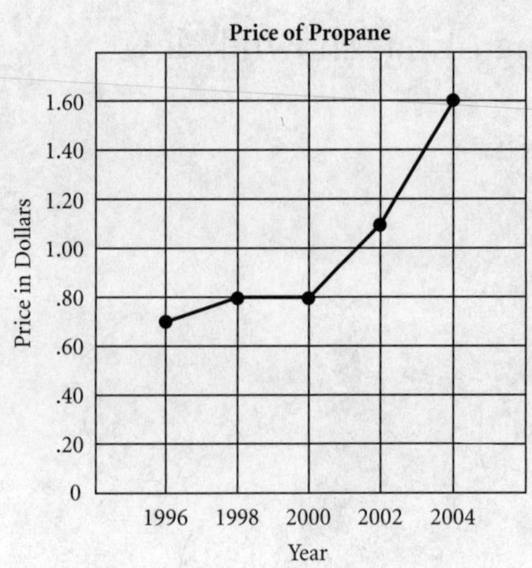

Based on this information, what can you predict will happen to the cost of propane in the next five years?

Ⓐ The cost will go up a lot.

Ⓑ The cost will go up a little.

Ⓒ The cost will go down a lot.

Ⓓ The cost will go down a little.

Standard 6, Key Idea 5

---

26 Newspapers, books, magazines, notepads, and more are all made from paper. Paper comes from trees. If you recycle paper, what will most likely occur?

Ⓐ Solid wastes will increase.

Ⓑ Water pollution will decrease.

Ⓒ Fewer trees will be cut down.

Ⓓ Paper mills will go out of business.

Standard 7, Key Idea 1

Name ______________________________

Date ______________________________

27 Examine the two electrical circuits shown below.

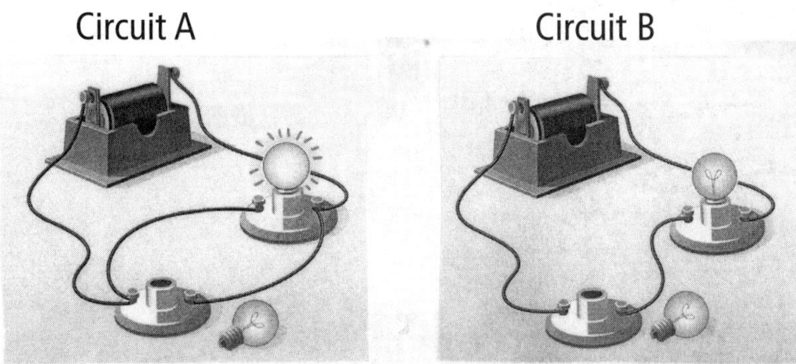

Explain why the bulb in Circuit A is lit but the bulb in Circuit B is not.

____

____

____

____

____

Standard 4: Physical Setting, 4.1e

Name ______________________________

Date ______________________________

28 The shape of each bird's beak indicates the type of food that bird eats. A toucan eats fruit. An eagle hunts small animals, and a heron spears fish.

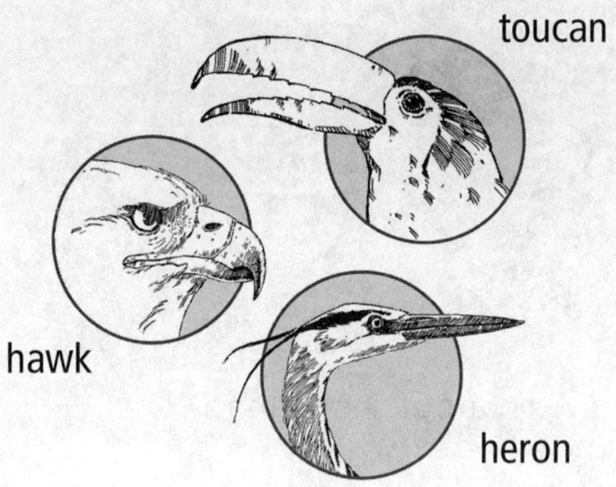

Is a bird's beak an inherited characteristic or an acquired characteristic? Explain your answer.

____

____

____

____

____

Standard 4: Living Environment, 2.1a

29 Luz did an experiment about the effects of an earthquake. She constructed buildings out of small blocks and placed them on each corner of a tray. The picture below shows the shapes of the buildings. Luz labeled the buildings: 1, 2, 3, and 4. Then she gently shook the tray back and forth three times to simulate an earthquake's motion. Buildings 2 and 3 quickly toppled.

How might the shape of a real building help prevent it from toppling during an earthquake?

____

____

____

Standard 4: Physical Setting, 2.1e

Name ______________________________

Date ______________________________

30 A student holds the two magnets shown below.

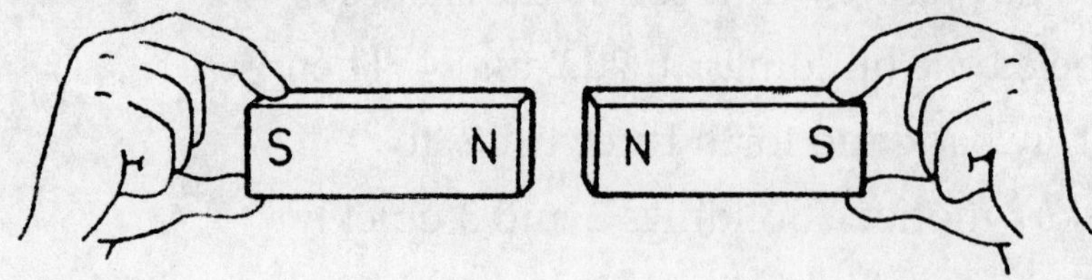

Describe how the student could increase the repulsion and how the student could increase the attraction between the magnets.

____

____

____

____

Standard 4: Physical Setting, 5.2b

Name ______________________________

Date ______________________________

31 Examine the positions of Earth, the moon, and the sun in the diagram below.

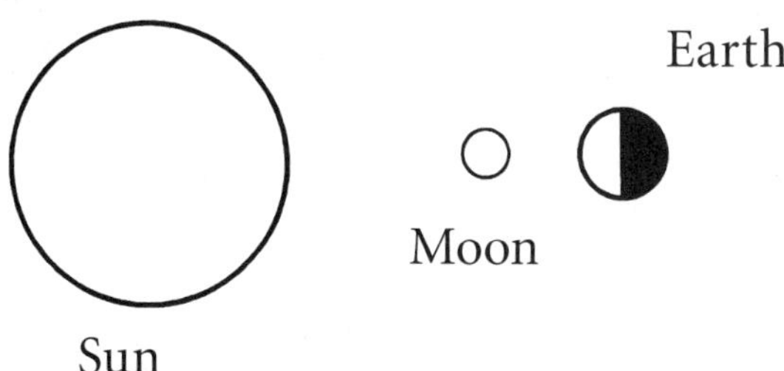

(a) Based on the diagram, which phase of the moon can be been from Earth?

____

____

(b) Which phase of the moon will be seen from Earth in about two weeks? In what positions will the Earth, the moon, and sun be then?

____

____

____

Standard 4: Physical Setting, 1.1a

Name ______________________________

Date ______________________________

32 Look at the picture of the eggs cooking on a stove.

(a) Think about how the eggs change as they cook. Describe some clues that show the eggs undergo chemical changes as they cook.

____

____

____

____

(b) Describe two other examples of how objects undergo chemical changes.

____

____

____

____

Standard 4: Physical Setting, 3.2c

Name ______________________

Date ______________________

1 Shoes have many features that make them safe and comfortable. Which feature can best prevent shoes from slipping on a smooth surface?

Ⓐ laces that keep the shoes tight

Ⓑ rubber on the sole of the shoe

Ⓒ padding on the inside of the shoe

Ⓓ soft material next to your foot

Standard 1, T1.1b

---

2 Lonny's teacher asked him to observe the crystal structure of table salt. Which tool should he use?

Ⓐ ruler

Ⓑ telescope

Ⓒ binoculars

Ⓓ hand lens

Standard 1, M3.1a

Name ______________________________

Date ______________________________

Practice Test 3

3 Devlin makes an electrical circuit using a copper wire, a battery, a switch, and a buzzer. When he tests the circuit, the sound of the buzzer is very low. How could he solve this problem?

Ⓐ remove the switch from the circuit

Ⓑ use a battery with a higher voltage

Ⓒ replace the buzzer with a light bulb

Ⓓ use a different metal for the wire

Standard 1, T1.5c

4 A class recorded the high temperature at their school each weekday for four weeks. The table shows their data.

Temperature (°C)					
Week	**Monday**	**Tuesday**	**Wednesday**	**Thursday**	**Friday**
1	20	24	25	23	23
2	24	24	22	25	24
3	23	24	26	24	27
4	22	23	25	26	27

During which week did the temperature increase each day?

Ⓐ Week 1

Ⓑ Week 2

Ⓒ Week 3

Ⓓ Week 4

Standard 1, M2.1b

Name ______________________

Date ______________________

5 Sue watches as her teacher uses a nail to hold up a paper clip, as shown below.

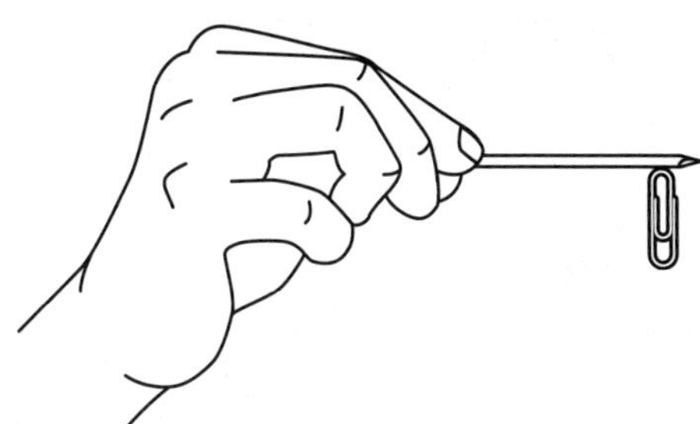

What is an observation Sue might record in her science journal?

Ⓐ Shaking the nail might make the paper clip fall.

Ⓑ The paper clip and the nail are magnetic.

Ⓒ The paper clip is attracted to the nail.

Ⓓ Another paper clip can be attached to the first one.

Standard 1, S1.1a

---

6 A student searches the Internet to learn how different instruments produce sounds. She reads the following statements. Which statement is an opinion?

Ⓐ Piccolos make sounds with a high pitch.

Ⓑ A snare drum sounds louder when it is hit harder.

Ⓒ French horns make deep, beautiful sounds.

Ⓓ A trombone's sound is changed by moving a slide.

Standard 2, Key Idea 3

Name ______________________________

Date ______________________________

Practice Test 3

7 The moon's phases go through a cycle. Part of the moon's cycle is shown below.

Which part of the moon's cycle is pictured?

Ⓐ new moon only

Ⓑ first-quarter moon only

Ⓒ waxing moon

Ⓓ waning moon

Standard 4: Physical Setting, 1.1a

---

8 Rachel and her family visited a city that was destroyed when it was covered by rocks, ash, and lava thousands of years ago. What most likely caused the city's destruction?

Ⓐ an earthquake

Ⓑ a tsunami

Ⓒ a volcanic eruption

Ⓓ weathering and erosion

Standard 4: Physical Setting, 2.1e

Name ______________________________

Date ______________________________

9 Study the weather instruments shown below.

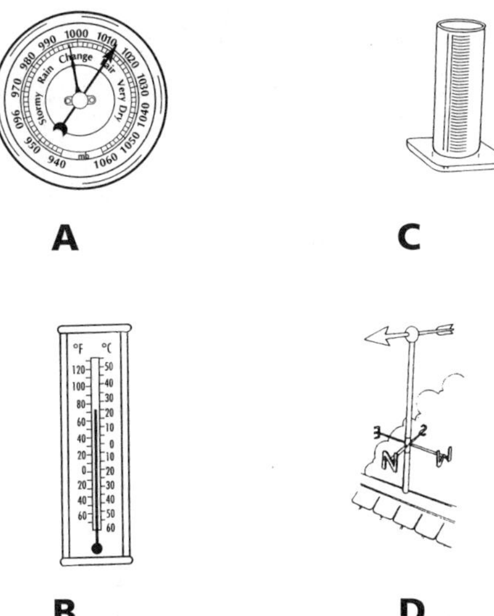

Which instrument measures wind speed?

Ⓐ instrument A

Ⓑ instrument B

Ⓒ instrument C

Ⓓ instrument D

Standard 4: Physical Setting, 2.1b

---

10 Lucia saw a canyon while on vacation. Which natural event most likely formed the canyon?

Ⓐ volcanic eruptions and lava

Ⓑ a tsunami

Ⓒ an earthquake

Ⓓ weathering and erosion

Standard 4: Physical Setting, 2.1d

11 This diagram shows the position of Earth as it orbits the sun.

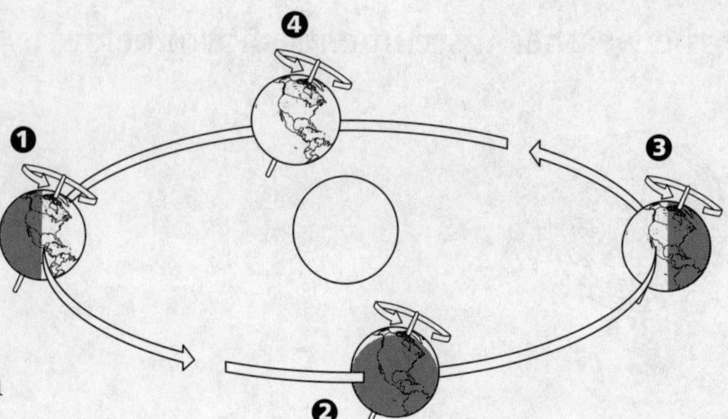

Which part of the diagram shows the position of the Northern Hemisphere during summer solstice?

Ⓐ 1

Ⓑ 2

Ⓒ 3

Ⓓ 4

Standard 4: Physical Setting, 1.1a

---

12 The illustration shows a student's school book.

What is the best estimate of the length of the book shown by the arrow?

Ⓐ 28 centimeters

Ⓑ 28 kilometers

Ⓒ 28 meters

Ⓓ 28 millimeters

Standard 4: Physical Setting, 3.1d

Name ______________________________

Date ______________________________

13 The picture below represents particles of a substance in a cup.

Which of the following is the best description of the substance?

Ⓐ It is a solid because the particles can move easily past each other.

Ⓑ It is a solid because the particles are not fixed in one place.

Ⓒ It is a liquid because the particles can move freely out of the cup.

Ⓓ It is a liquid because the particles can move around each other.

Standard 4: Physical Setting, 3.2a

---

14 A magnet will attract which of the following materials?

Ⓐ iron and steel

Ⓑ aluminum and iron

Ⓒ copper and aluminum

Ⓓ steel and copper

Standard 4: Physical Setting, 5.1e

Name ______________________________

Date ______________________________

15 Look at the fishing rod below.

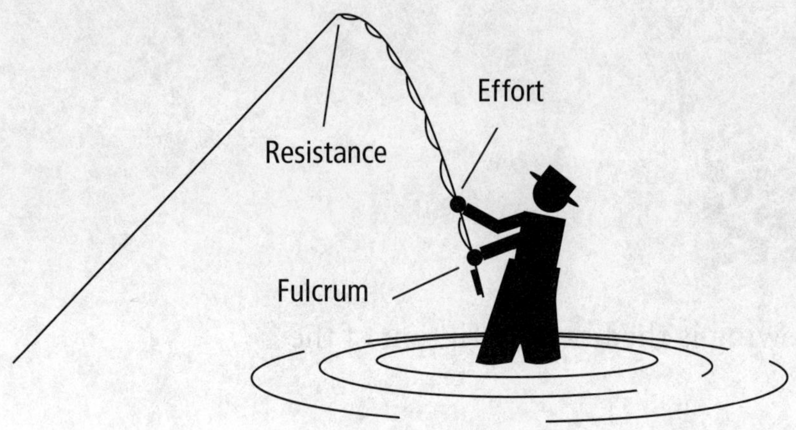

What type of simple machine is the fishing rod?

Ⓐ a lever

Ⓑ a pulley

Ⓒ an inclined plane

Ⓓ a wheel-and-axle

Standard 4: Physical Setting, 5.1f

---

16 Which color is best to wear in sunlight during the summer if you want to stay cool?

Ⓐ black

Ⓑ white

Ⓒ red

Ⓓ blue

Standard 4: Physical Setting: 4.1d

Name ______________________________

Date ______________________________

17 Maria's earlobes are attached to her neck. Her father's earlobes hang freely, but her mother's earlobes look like Maria's. What is this characteristic of Maria's earlobes called?

Ⓐ a trait

Ⓑ heredity

Ⓒ a gene

Ⓓ a development

Standard 4: Living Environment, 2.2a

18 Look at the bar graph below.

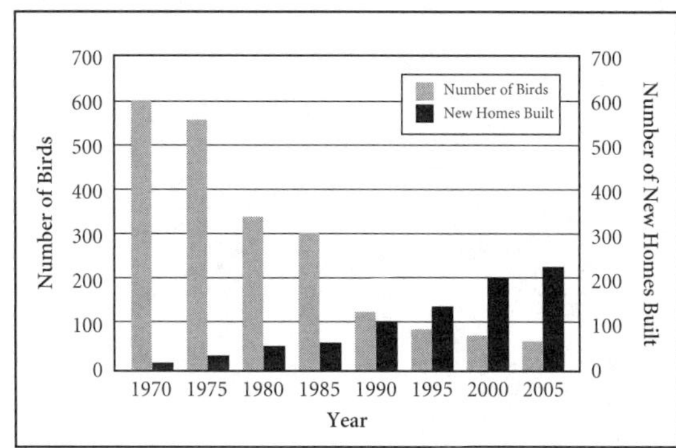

Which statement is the best interpretation of the data in the bar graph?

Ⓐ Soil erosion caused the number of birds to decrease.

Ⓑ High temperatures caused the birds to migrate north.

Ⓒ An increase in home building caused the number of birds to decrease.

Ⓓ A decrease in the number of birds caused more new homes to be built.

Standard 4: Living Environment, 7.1c

Name ______________________________

Date ______________________________

19 The picture below shows a desert environment.

Which item in the picture is nonliving?

Ⓐ cactus

Ⓑ rabbit

Ⓒ rock

Ⓓ snake

Standard 4: Living Environment, 1.1c

20 Which of the following list of items do animals need in order to live?

Ⓐ air, water, and food

Ⓑ soil, sunlight, and water

Ⓒ wood, rocks, and rain

Ⓓ food, shelter, and clothing

Standard 4: Living Environment, 1.1a

Name ______________________________

Date ______________________________

21 What is the function of a tree's roots?

Ⓐ to digest food

Ⓑ to produce new trees

Ⓒ to use sunlight to make food

Ⓓ to take in water and nutrients

Standard 4: Living Environment, 3.1b

---

22 What is a human called at the beginning of its life cycle?

Ⓐ an adolescent

Ⓑ an adult

Ⓒ an infant

Ⓓ a teenager

Standard 4: Living Environment, 4.1f

Name ______________________________

Date ______________________________

23 A house plant is placed near a window. In which direction will the plant most likely grow?

Ⓐ The plant will grow towards the window.

Ⓑ The plant will grow away from the window.

Ⓒ The plant will only grow upwards.

Ⓓ The plant will not grow at all.

Standard 4: Living Environment, 5.2a

---

24 Rex is a puppy. Every week Rex is weighed and measured. Each week he weighs a little more and is a little bigger. What process is Rex undergoing?

Ⓐ digestion

Ⓑ growth

Ⓒ hibernation

Ⓓ reproduction

Standard 4: Living Environment, 4.2a

Name ______________________________

Date ______________________________

25 Stars of different colors have different temperatures, as shown in the table below.

Color of Star	Temperature of Star (°K)
Blue	25,000
White	6,000
Yellow	6,000
Orange	3,500
Red	3,000

Which kind of graph would best show this information?

Ⓐ bar graph

Ⓑ picture graph

Ⓒ circle graph

Ⓓ scatter plot

Standard 6, Key Idea 2

---

26 Judy lights a candle and lets it burn for 1 hour. She then blows out the candle. She decides to record the properties of the candle. What property can she measure using a balance?

Ⓐ color

Ⓑ length

Ⓒ mass

Ⓓ shape

Standard 6, Key Idea 4

Name ______________________

Date ______________________

27 The picture below shows two students picking up pieces of a broken glass beaker.

Identify one thing you can observe in the picture and one thing you can infer based on the picture.

____

____

____

____

Standard 7, Key Idea 2

Name ______________________________

Date ______________________________

28 The hair dryer shown below uses electricity as a form of energy. If you turn on the hair dryer, this energy changes to other forms of energy.

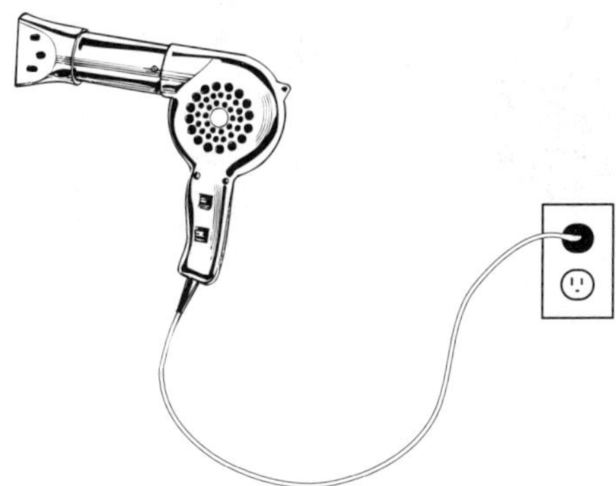

Identify two types of energy transformations that occur when you turn on the hair dryer.

____

____

____

Standard 4: Physical Setting, 4.2b

29 How are living things different from nonliving things?

____

____

____

____

____

Standard 4: Living Environment, 1.2a

Name ______________________

Date ______________________

30 Look at the diagram of the water cycle below.

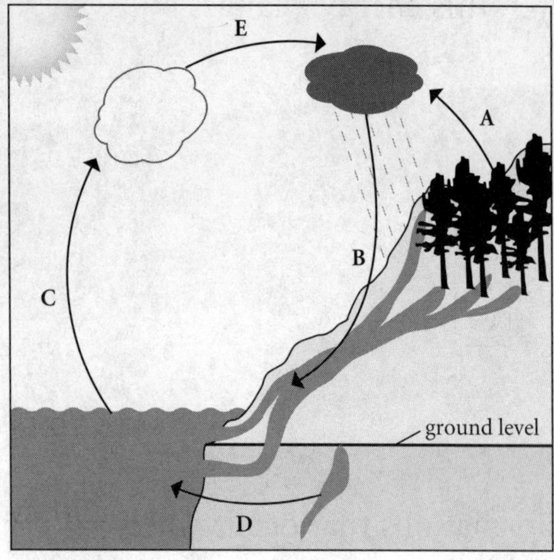

At what point in the water cycle does evaporation occur? Explain this process.

____

____

____

____

____

Standard 4: Physical Setting, 2.1c

Name ______________________________

Date ______________________________

31 The speed of an object is the distance it travels divided by the time required to travel that distance. A student uses the following setup to compare the speeds of several toy cars as they roll down a ramp.

(a) Write steps the student could follow in his investigation.

____

____

____

____

____

(b) Describe a good way in which the student could record his data.

____

____

____

Standard 1, S2.1a

Name ______________________

Date ______________________

32 The organisms in this food chain depend on one another for food.

(a) An important organism is missing from the food chain. Identify it as either a producer or a consumer, and explain your reasoning. Then identify where the missing organism should be positioned in the food chain.

____

____

____

____

(b) Describe how energy flows among the organisms in the food chain. Include the missing organism in your description.

____

____

____

____

____

____

Standard 4: Living Environment, 6.1c